Made in United States
Cleveland, OH
05 November 2026

47662564R00054